How to Make Patent Drawings

A Brief Treatise on Patent Drafting for the Use of
Students, Draftsmen and Inventors

By
L. H. FULMER

British Library Cataloguing-in-Publication Data
A catalogue record for this book is available from the
British Library

Technical Drawing and Drafting

Technical drawing, also known as 'drafting' or 'draughting', is the act and discipline of composing plans that visually communicate how something functions or is to be constructed.

It is essential for communicating ideas in industry, architecture and engineering. The need for precise communication in the preparation of a functional document distinguishes technical drawing from the expressive drawing of the visual arts. Whereas artistic drawings are subjectively interpreted, with multiply determined meanings, technical drawings generally have only one intended meaning. To make the drawings easier to understand, practitioners use familiar symbols, perspectives, units of measurement, notation systems, visual styles, and page layout. Together, such conventions constitute a visual language, and help to ensure that the drawing is unambiguous and relatively easy to understand.

There are many methods of constructing a technical drawing, and most simple among them is a sketch. A sketch is a quickly executed, freehand drawing that is not intended as a finished work. In general, sketching is a quick way to record an idea for later use, and architects sketches in particular (in a very similar manner to fine artists) serve as a way to try out different ideas and establish a composition before undertaking more finished work. Architects drawings can also be used to convince clients of the merits of a design, to enable a building constructer to use them, and as a record of completed work. In a similar manner to engineering (and all other technical drawings), there is a set of conventions (i.e particular views, measurements, scales, and cross-referencing) that are utilised.

As opposed to free-sketching, technical drawings usually utilise various manuals and instruments. The basic drafting procedure is to place a piece of paper (or other material) on a smooth surface with right-angle corners and straight sides – typically a drawing board. A sliding straightedge known as a 'T-square' is then placed on one of the sides, allowing it to be slid across the side of the table, and over the surface of the paper. Parallel lines can be drawn simply by moving the T-square and running a pencil along the edge, as well as holding devices such as set squares or triangles. Other tools can be used to draw curves and circles, and primary among these are the compasses, used for drawing simple arcs and circles. Drafting templates are also utilised in cases where the drafter has to create recurring objects in a drawing – a massive time-saving development.

This basic drafting system requires an accurate table and constant attention to the positioning of the tools. A common error is to allow the triangles to push the top of the T-square down slightly, thereby throwing off all the angles. Even tasks as simple as drawing two angled lines meeting at a point require a number of moves of the T-square and triangles, and in general drafting this can be a time consuming process. In addition to the mastery of the mechanics of drawing lines, arcs, circles (and text) onto a piece of paper – the drafting effort requires a thorough understanding of geometry, trigonometry and spatial comprehension. In all cases, it demands precision and accuracy, and attention to detail.

Conventionally, drawings were made in ink on paper or a similar material, and any copies required had to be laboriously made by hand. The twentieth century saw a shift to drawing on tracing paper, so that mechanical copies could be run off efficiently. This was a substantial development in the drafting process – only eclipsed in the twenty-first century with 'computer-aided-drawing' systems (CAD). Although classical draftsmen and women are still in high demand, the mechanics of the drafting task have largely been automated and accelerated through the use of such systems. The development of the computer had a major impact on the methods used to design and create technical drawings, making manual drawing almost obsolete, and opening up new possibilities of form using organic shapes and complex geometry.

Today, there are two types of computer-aided design systems used for the production of technical drawings; two dimensions ('2D') and three dimensions ('3D'). 2D CAD systems such as AutoCAD or MicroStation have largely replaced the paper drawing discipline. Lines, circles, arcs and curves are all created within the software. It is down to the technical drawing skill of the user to produce the drawing – though this method does allow for the making of numerous revisions, and modifications of original designs. 3D CAD systems such as Autodesk Inventor or SolidWorks first produce the geometry of the part, and the technical drawing comes from user defined views of the part. This means there is little scope for error once the parameters have been set.

Buildings, Aircraft, ships and cars are now all modelled, assembled and checked in 3D before technical drawings are released for manufacture.

Technical drawing is a skill that is essential for so many industries and endeavours, allowing complex ideas and designs to become reality. It is hoped that the current reader enjoys this book on the subject.

PREFACE

A great many books have been written on drawing, both mechanical and architectural, but to my knowledge very few, if any, have ever given information in regard to the making of Patent drawings, excepting of course the U. S. " Rules of Practice." In this progressive period, when there are so many inventions being discovered and so many patents being secured to protect the same, there is naturally a demand for good Patent Draftsmen. As every patent issued requires drawings, and the U. S. Patent Office will accept only drawings which are made by men skilled in the art, and made under certain rules, I think that a book disclosing information on this subject will be very useful, not only to the man wishing to fit himself to hold a position as draftsman, but also to the inventor who desires to make his own drawings, either to save himself the expense, or because he is so located as to be unable to secure the services of a good man.

CONTENTS

PREFACE

CHAPTER   . PAGE

   I. Instruments and Materials........   5

  II. Use of Instruments................. 13

 III. Penciling the Drawing.............. 18

 IV. Inking the Drawing...............   25

   V. Sections and Section Lining............. ... 28

 VI. Outline Shading........................ 31

VII. Surface Shading.......................... 34

VIII. Lettering the Drawing.................... 36

 IX. Sketching................................. 37

   X. Care of Instruments...................... 39

 XI. Blue Printing, etc....................... 41

XII. Useful Hints............................. 43

CHAPTER I

Instruments and Materials

Introduction.—In this chapter I have listed instruments and materials necessary to make a first class Patent drawing, and they should be secured before attempting any work. They can be purchased at most Art stores, or if desired, can be ordered through the mail from firms dealing exclusively in draftsman's supplies. To the man just starting this work the instruments may seem an expensive investment, but if good instruments are bought this is not so, as they will last a lifetime, and the draftsman is always sure of securing the best results in his work. The beginner should never hinder himself by using instruments of an inferior grade, as the accuracy and speed of patent drawings as well as the finish depends to a great extent on the quality of the instruments.

The instruments and materials I have divided into two lists, the first consisting of those which are absolutely necessary, and the second those not necessary but good to have for certain work.

The first list consists of drawing board, drawing paper, thumb tacks, pencil, ink, erasers (pencil and ink), tee-square, 30° and 45° triangles, scale, irregular curve, compass (with pen and pencil point, and lengthening bar), dividers, bow dividers, bow pencil, bow pen, and ruling pen (medium size).

In Plate 1 (2) illustrates drawing board, (3) drawing paper, (4) thumb tack, (5) 45° triangle, (6) 30° triangle, and (7) tee-square.

The second list consists of proportional dividers, erasing shield, crow quill pen, transfer sheet, stylus, pencil pointer, burnisher, cleaning eraser, and a small ruling pen.

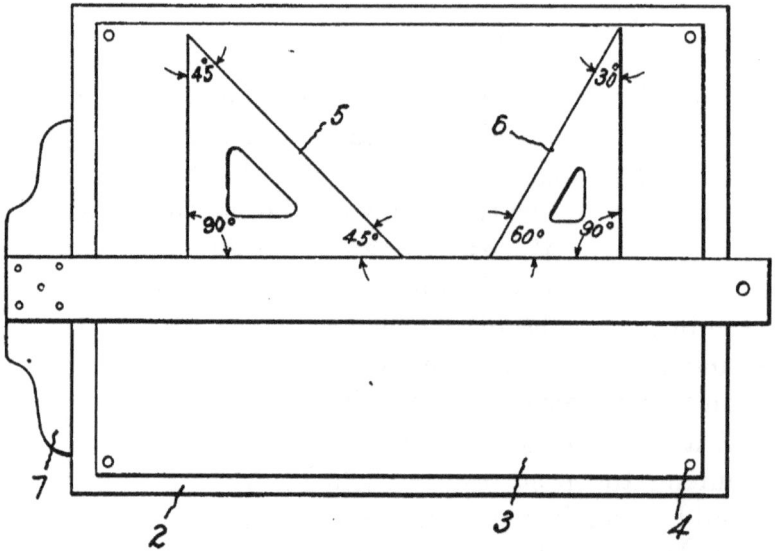

PLATE 1

Drawing Board.—The drawing board is a rectangular board on which is fastened the paper. Usually it is made of white pine about ¾″ or ⅞″ thick, and with the grain running the long ways of the board. The most convenient size for patent work is 13″ x 18″. It must never be less than 10½″ x 15½″. The front face, that is, the working surface, should be smooth and perfectly flat. The four edges of the board should form a perfect rectangle, or at least the adjacent edges at the lower left hand corner should be at right angles to each other. It is generally used with the short sides at top and bottom.

Drawing Paper.—The paper must be pure white with a smooth calendered surface and must have a thickness to equal three-ply Bristol board. Of course the better the quality the more easily erasures may be made. The size of the sheets must be 10″ x 15″. For patent work I recommend Reynolds Bristol board of three ply, which comes in sheets about 12⅜″ x 15¼″, allowing a margin outside the cutting lines of the drawing on which you can try your pens when inking. Patent firms usually furnish their workmen with paper.

Thumb Tacks.—Thumb tacks are used to fasten the sheet of paper to the drawing board. Those made of one piece of hard steel with a flat head and fine needle point are the best. They are very inexpensive and a dozen of the best quality should be secured.

Pencil.—The drawing pencil for laying out all drawings should be of good quality; not too hard as it will make indentures in the surface of the paper, and in case of an erasure cannot be removed; nor should it be too soft, as it will wear away quickly and it would be impossible to secure sharp lines. I recommend a pencil of HHHH hardness. This should be sharpened to a long conical point secured by holding the lead at an angle to a piece of sandpaper or emery cloth and drawing it back and forth at the same time revolving the same. While penciling the drawing the point should be watched and when it becomes rounded or blunt should be touched up on the sandpaper.

The pencil should be held vertical or nearly so and should be pressed on the paper lightly. If inclined at all it should be in the direction in which it is moved.

Ink.—India waterproof ink is the best for this work, and I recommend that put on the market by Higgins. This ink flows freely and presents a perfectly black line. It comes in a bottle with a short quill in the stopper. By means of this quill ink is easily placed between the blades of the pens and compasses. Be sure before using to shake the bottle well as it will settle and is liable to be thick near the bottom. If it should become too thick to flow properly it can be thinned by adding a few drops of ammonia.

Erasers.—For a pencil eraser a soft pliable rubber should be used. Do not bear on the paper but rub softly. Pressure on the rubber does not remove the line any sooner.

For ink eraser, use a stiff rubber with very little grit in it. For most work a pencil eraser will answer the purpose. It will take a little longer, but the surface of the paper will not be injured as it might by a hard pressure on the ink eraser.

Tee-Square.—The tee-square is used as an edge on which to draw horizontal lines; and as a surface to place the triangles against for drawing straight lines in oblique and vertical directions. This instrument derives its name from its shape which is that of the letter " T." It consists of two parts: the blade, and the stock. The horizontal part of the letter "T" is the stock and the vertical part the blade. They are joined together at right angles to one another. The stock is placed against the working edge of the board and the lines are drawn along the edge of the blade. The tee-square for this work should be 18″ long and preferably of wood with celluloid edges.

To use, grasp the tee-square near the center of head with the left hand, and slide the inner edge of the head against the edge of the board, always making sure the edge is in perfect contact, otherwise the lines would not be parallel.

30° and 45° Triangles.—Triangles are used as a guide for the pencil or ruling pen in drawing lines at an angle to the tee-square. They are always used to make section lines. Those made of celluloid are to be most recommended for this work, as the draftsman is able to see through the transparent material and thus work to better advantage. There are two triangles needed in this work, one known as the 45° and the other the 30°. They both have one right angle, and in addition the one has two 45° angles, the other one 30° angle and one 60° angle. The most convenient size for this work are those with legs about 6″ to 8″ long.

To use, place triangle against the upper edge of tee-square and hold in perfect contact with the same by the first three fingers of left hand, while the rest of the hand holds the tee-square firmly.

Scale.—The scale is used principally to mark off accurately the cutting out lines and the border lines of the drawing. It is also used in cases in which the draftsman has a model or blue prints from which he wishes to secure dimensions or from which he wishes to make a drawing at a reduced scale. The 12″ boxwood scale is the best for this work.

Irregular Curve.—Irregular curve (sometimes called French curve) is used to draw curved lines that cannot be made conveniently with a compass or other instruments. They are of wood, hard rubber or celluloid. Celluloid is the best. The curve to be laid out is usually constructed by locating a number of predetermined points and then sketching in free hand a curve passing through these points. Then the irregular curve is adjusted to follow as closely as possible this free hand curve. By continually applying the irregular curve to pass through the points the desired curve can be secured. When using the irregular curve the ruling pen must always be held tangent to the surface of the curve and be held vertical. Be careful not to press against the edge of the curve hard enough to

PLATE 2

press the blades of the pen together. Plate 2 shows an irregular curve.

OTHER MATERIALS AND INSTRUMENTS

Proportional Dividers.—These differ from ordinary dividers in that they have four points, two of which when set to the full dimensions will be reproduced by the other pair, but at a different scale. In appearance they are very much like a pair of double ended dividers. They are provided with a movable slide which can be fastened when the ends of the instrument are adjusted for the desired reduction. They are generally used for transferring distances from one place to another and either reducing or enlarging the scale in the operation. As the draftsman has to show on the small sheet of paper provided for this work machines, etc., of all sizes, it can be readily seen that this instrument comes in very handy, especially if working from a large model or blue prints. Some dividers are marked for the proportions and are provided with a rack movement for adjustment and with a screw for holding in the

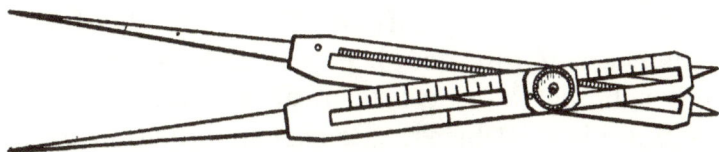

PLATE 3

desired position. This instrument is rather expensive to buy but the draftsman will usually find the firm by whom he is employed has a pair for the use of their men. Plate 3 shows a pair of proportional dividers.

Erasing Shield.—Used for erasing certain lines without disturbing adjacent lines. It is made of a thin piece of metal with different size slots in it. The shield is placed on the line to be erased with the slot uncovering the line. Then by rubbing the eraser in the slot the line is erased without erasing any other line.

Crow Quill Pen.—This is a pen similar to an ordinary writing pen but very much finer. It is used to ink small parts which require very fine and delicate lines, and which cannot be made with instruments.

Transfer Sheet.—A transfer sheet is a sheet of thin, tough paper one side of which is blackened with graphite. It is used when it is desired to make an exact copy of a drawing, print, sketch, etc.

To use, place the transfer sheet with the graphite side down against your Bristol board. On the transfer sheet place the drawing, or whatever is to be copied, in the exact position desired, and fasten with thumb tacks, being sure to place the tacks outside the cut out lines of your drawing. Then with a stylus (described later), or a very hard pencil, follow carefully the lines of the drawing, and the impress will make a clear, clean-cut copy on your bristol board. You can easily make your own transfer sheet by scraping or filing the graphite from a very soft pencil on one side of a plain white sheet of thin, tough paper, and then with a soft cloth rubbing the graphite until the entire surface of the paper is a jet black. Be sure and dust off all loose particles so as not to smear the surface of the bristol board. Do not use typewriter carbon sheets for if a mistake is made it is impossible to erase the marks.

Stylus.—A stylus is an instrument with a hard sharp steel point which is used in conjunction with a transfer sheet to make exact copies of prints, etc. It is held in the hand the same as a pencil or pen, and by pressing the sharp steel point on the lines of the print, the copy is made.

Pencil Pointer.—A pencil pointer is used to secure very sharp points on the leads in pencils and instruments. It consists of a piece of fine sandpaper, or emery cloth, fastened to a flat board, provided with a handle.

Burnisher.—A burnisher is used to polish the surface of the bristol board when it has been made rough by an erasure. It usually consists of a steel or glass rod which has an end or knob with a highly polished surface. This is rubbed over the rough surface of the paper until a hard, smooth surface is secured by the friction.

Cleaning Erasers.—Cleaning erasers, for cleaning the lint, dirt, etc., from the surface of the drawing paper, are very convenient to have. The soft sponge rubber, or the art gum, are the best for this work.

CHAPTER II

Use of Instruments

Introductory.—The following instruments usually
come in sets in a case or box, and will therefore be considered under a separate chapter. However, it is not
necessary for a beginner to always buy a case of instruments as some sets contain certain pieces that are
seldom if ever used, and then sometimes separate instruments can be bought of just as good quality, but
much less in price. If possible it might be wise before
purchasing to secure the advice of some draftsman who
is experienced in the use and care of instruments.

I will describe the use and construction of the instruments which I consider necessary for first-class
patent work.

Ruling Pen.—The ruling pen is one of the most important instruments which a draftsman uses. On its
skillful use depends the good looks of the drawing. It
is used to draw straight lines along the tee-square
blade, or the edges of the triangle. It has two steel
blades which are opened and closed by an adjusting
screw. These blades should be tempered properly, so
that they will not be soft and wear away quickly. Ink
is inserted between the blades by means of a quill which
will be found in the ink bottle. Never dip the blades
in the ink. After filling the pen with ink test it on
the edge of your paper to see if you have the desired
width of line. By regulating the distance between the
blades with the adjusting screw, a line of any desired

thickness can be secured. Hold the pen in the right
hand (unless you are left handed) with the thumb and
the first and second fingers, with the adjusting screw
away from you, and place the pen against the tee-
square blade or triangle with its blades parallel to the
direction in which you draw the line. Hold the pen
nearly perpendicular to the surface of the drawing,
inclining very slightly in the direction in which the
line is to be inked. The other fingers rest lightly on
the tee-square or triangle. Hold the pen at the start-

RULING PEN (LARGE)   LENGTHENING BAR   BOW PENCIL

RULING PEN (SMALL)   BOX OF LEADS   BOW DIVIDERS

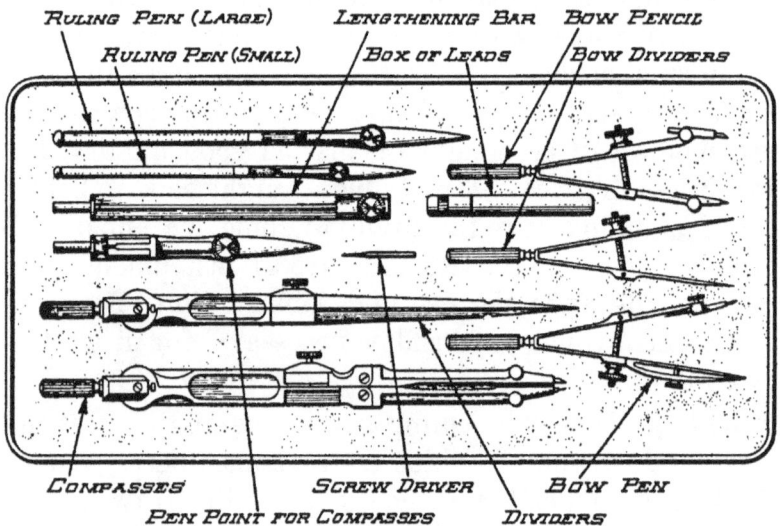

COMPASSES   SCREW DRIVER   BOW PEN

PEN POINT FOR COMPASSES   DIVIDERS

PLATE 4

ing point just an instant until the ink begins to flow,
then move at an even speed to the right until the desired length of line is made. The pen should be removed immediately on reaching the end of the line,
otherwise the line is liable to spread. If a number of
lines are to meet at a point, all lines should be drawn
from the point and not towards it. Be sure you do
not press the blades of the pen too hard against the

edge of the tee-square or triangle, as this will cause an uneven line. The pen should touch the edge lightly and the pressure on the pen should be uniform throughout the length of the line, otherwise it will cause an unevenness of the line. Do not lay away your pen after inking without first thoroughly cleaning. A good many draftsmen have two sizes of pens, a large one for the thick shade and border lines, and a small one for fine shade lines. This is not necessary, however, as a medium size pen will answer all purposes. Plate 4 shows a set with two pens.

Compass.—The compass is used to draw circles or parts of circles with either pencil or ink. It has a fixed leg with a needle point, used as a centre around which the circle is drawn, and another leg with removable pencil and pen points. The legs are jointed so that the points can be made perpendicular to the paper when large circles are drawn. For circles larger than can be conveniently made with the compass the draftsman should use the extension bar which is always furnished. This bar extends the pencil or pen points so that circles of considerable radii can be made. The two legs of the compass are joined in what is called the head of the compass. This head should be well considered when buying this instrument. It should be constructed so that the legs are held firmly in any position, but not so tight that they cannot be readily adjusted. The heads of some compasses are fitted with small slots into which fits a key for tightening the same. Others are held firmly with a small set screw which can be tightened with the screw driver, which is always furnished with the sets of instruments. The pencil leg of the compass should have hard lead, sharpened to a conical point. For ink work remove the pencil leg by loosening the little set screw and insert the pen leg. The pen point of the pen leg is adjusted for thickness of lines the same as the ruling pen. Always insert the ink between the blades with the quill as previously described. The needle point in the fixed leg is always a separate piece of steel, held in a socket in the end of the leg by a small set screw. This steel piece has a square shoulder below which projects a very fine needle like point.

To open the compasses to a desired radius, hold with the needle point leg resting between thumb and fourth finger, and the other leg between the two middle fingers.

To scribe a circle or arc, hold the compasses lightly at the top, between the thumb, first and second fingers, and rotate from right to left. Always press lightly on the needle point, just enough to keep it from slipping. Be careful not to dig a hole in the paper.

Only one hand should be used to operate the compasses. The beginner has a great tendency to use both hands, which makes him appear awkward and gives a bad impression although he may be a good workman. The only exception is when the extension bar is used, then it is absolutely necessary to use both hands to hold the point perfectly steady.

One style of compass with pencil point adjusted is shown on Plate 4, also the pen point and lengthening bar used with the same.

Dividers.—This instrument is used to transfer distances or spaces from one drawing to another or from one part of a drawing to another. Also used for dividing circles, arcs and lines into equal parts. It is simliar in design to the compass, but the legs are fixed. There are no joints and both legs have long tapered points very sharp, so as not to punch large holes in the paper. It should be held while using the same as the compass, but when spacing turn the instrument alternately from right to left. See Plate 4 for the style of dividers most commonly used.

Bow-Dividers.—This instrument is used for the same purpose as dividers, but is much more convenient and accurate for small work. It has an adjusting screw which will allow for very fine adjustment. Some of these instruments are constructed with a single threaded piece fastened to one leg and adjusted by a screw on the outside of the other leg. This type is shown on Plate 4. Others are fitted with a piece with both right and left-hand threads which moves the legs of the instrument by turning a central thumb screw. This latter type is used more for the reason that it can be easily adjusted with one hand. When spacing the instrument is held in the hand the same as the large dividers and also used in the same manner.

Bow-Pencil.—A bow-pencil is a pencil compass for making small circles and arcs and should be used for all small work in preference to compasses. The pencil point should be fitted with hard lead sharpened to a conical point. The means of adjustment is the same as the bow-dividers.

Bow-Pen.—This instrument is used to ink all circles and arcs penciled with the bow-pencil. It is constructed the same as the bow-pencil except that it has a pen instead of a pencil point. Do not ink circles with this pen which require the pen to be opened to such an extent as to cause the point to make a ragged line. Always use large compasses for circles over one inch or one inch and one-quarter radius. Ink is inserted in pen point, and the desired thickness of a line is secured in the same manner as described under ruling pen.

CHAPTER III

PENCILING THE DRAWING

A drawing poorly penciled is seldom well inked. Therefore, the beginner should be very particular about his pencil work and before he starts should be sure his pencil has a sharp, conical point and that the leads in his other instruments are also well sharpened. Care should be taken to keep them in this condition as this is a very important requisite in the execution of this part of the work. As the point wears away it should be touched up frequently on a piece of emery cloth or sandpaper.

When drawing a line the pencil should always be held vertical, or nearly so, and should be pressed very lightly on the paper, making the lines as fine as is consistent with clearness. If the pencil is inclined at all it should be in the direction in which it is moved.

With the pencil and instruments in perfect working order, the next important step is to place the drawing paper on the board and fasten it so that the edges are parallel to the edges of the board. First lay the bristol board flat on the board; put on the tee-square with the head at the left-hand side of the board; then slide tee-square up nearly to the top and arrange the paper level with the blade; with the right hand hold the paper still, sliding the tee-square down a little, and then pin the top of the paper with thumb tacks about ¼" from each corner. Next, pin the lower corners, being sure to smooth the paper out perfectly flat.

Now, the border lines and cutting out lines should be laid out accurately with the scale, to the dimensions given on Plate 5. The draftsman should be very particular to get these dimensions exact, as the Patent Office will refuse to accept drawings unless they are the size here specified.

The cutting out lines, that is, the outside lines A, B, C, D, where the drawing is cut out when finished must be 10″ x 15″.

The border lines a, b, c, d, within which all drawing must be made is 8″ x 13″. This leaves an inch margin all around.

The shorter side of the drawing is considered the top, that is, side "A," "B."

A light pencil line e. f. (don't ink) should be drawn 1¼″ below the line a, b. In this space, a-b-e-f, no marks of any kind should be made. This space is reserved for putting in title, etc., at the Patent Office in Washington.

In the lower right-hand corner should be penciled the words "Inventor," "By" and "His Attorney," as shown in Plate 5. The inventor's name, if known, can also be printed just below the word "Inventor." There is no fixed rule as to the distance from lower edge to place the printed words, but enough space should be allowed between the inventor's name and the words "His Attorney" so that the attorneys will be able to sign their names. About 1½″ usually is allowed.

There is also no fixed rule as to the style or size of the lettering to be used. However, it is preferable to always use the same size and style.

The sheet of bristol board is now in shape to start penciling the drawing. Before making a line, however, the draftsman should thoroughly study the model, sketches, prints or whatever information has been given him from which to make his drawings. He should be sure he thoroughly understands the invention, and then decide on the different views which he thinks will be necessary to fully illustrate the same. There is always a choice of views and those should be made which convey the greatest amount of information, consistent with clearness. Generally three views of an object are all that are necessary to show its form and its use. If, however, its construction is complicated, it may require sectional views, detail views, or even perspective views in addition to the general views. The draftsman should imagine himself reading the drawings and then decide on how many views are necessary to give complete information in a clear and concise manner. The draftsman must keep in mind that he must show clearly every feature of the invention which the inventor wishes to protect with his patent.

When the invention is an improvement on some old machine, the drawing must show in one view, at least, so much of the old structure as will suffice to show the connection of the invention therewith. Then also must be shown separate views of the device detached from the old structure.

After deciding on the views to be shown the draftsman should make a study as to how the different views should be arranged, and how many sheets of drawings will be necessary. The views should of course be arranged systematically on the sheet in order that the space may be used to the best advantage, and the drawing as a whole will present a pleasing appearance.

The scale to which the drawing will have to be drawn will have to be given considerable thought, as the Patent Office does not want any more sheets used than is absolutely necessary, yet, on the other hand, it must be large enough so as not to have the mechanism crowded, or to have it drawn to such a small scale as not to be easily read.

The three views that will fully illustrate most devices, unless of course they are of a complicated mechanism, some parts of which are invisible, are plan view, front view, and side view.

The drawing which represents the object as if it were resting on a horizontal plane and the observer looking at it from directly above, is called a plan or top view.

A front view (or front elevation) is drawn so that it represents a view with the observer looking directly at the front of the object.

The side view or side elevation shows the observer's line of vision the same as in front view, but the object is shown as if revolved on its axes 90˚ either to right or left.

Having decided on the views desired on the sheet and calculated as accurately as possible the amount of space each view will require, and the exact place it should be located on the sheet, you can now proceed to pencil the views.

As most views centre around certain points, it is best to begin by locating these points and drawing through them centre lines both vertical and horizontal, carefully locating all your distances thereafter from these lines.

It is best usually to pencil the front elevation first, but, of course, if it is possible by projection to work on more than one view, do so, as much time is saved in this way.

After the general views are made you can then proceed with the detail and sectional views if such are necessary. If a sectional view is made always, if pos-

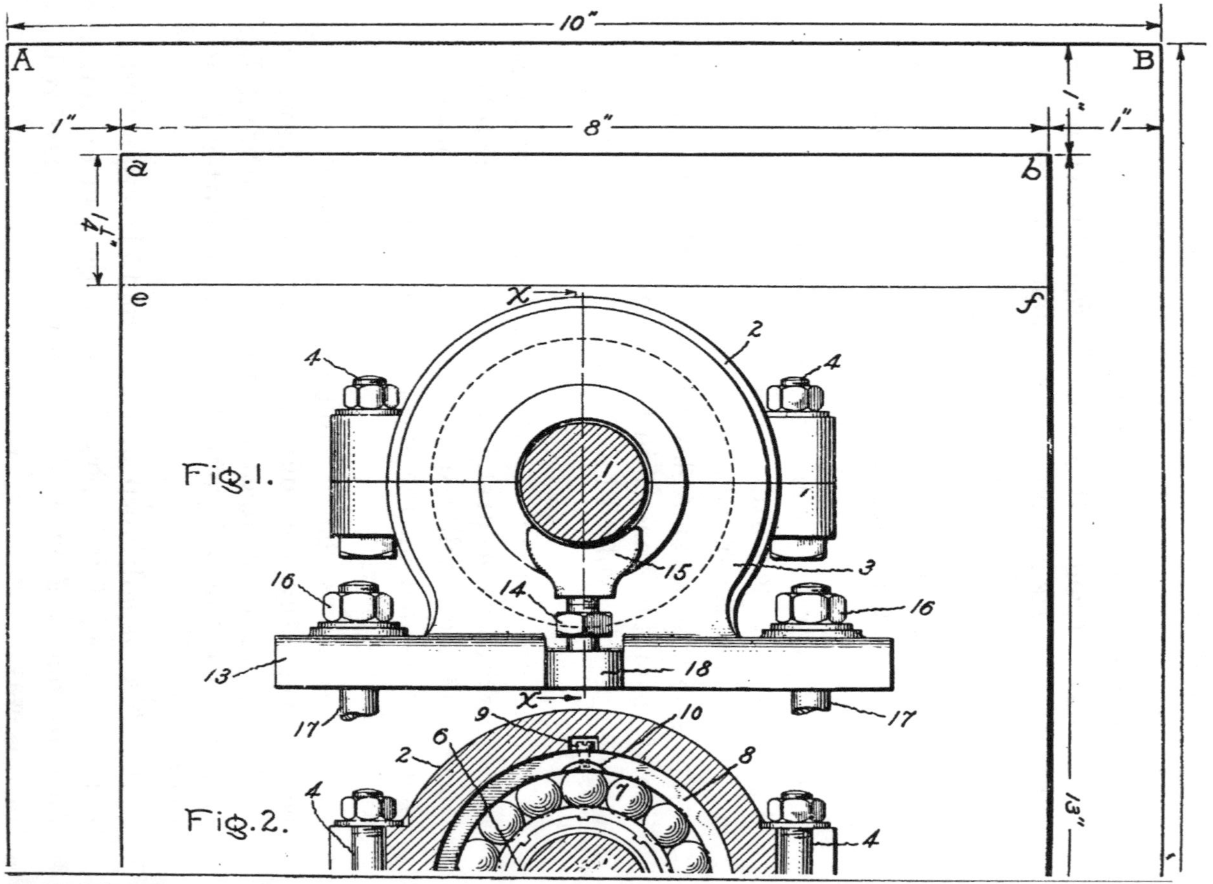

Fig.1.

Fig.2.

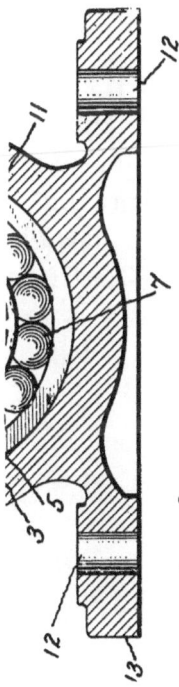
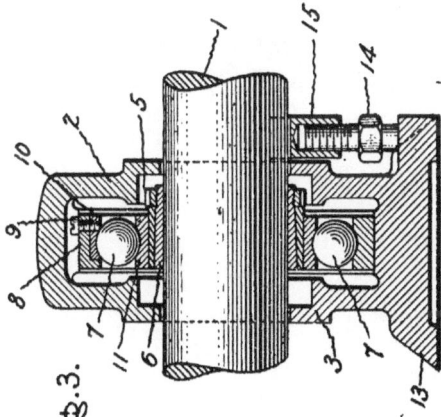

Fig.3.

Fig.4.

ABT. 1½"

1"

Inventor:
John W. Doe,

by

His Attorney.c

sible, show where the section is taken with a broken line as at X-X in Fig. 1, Plate 5.

Perspective views are used by some draftsmen for showing up certain objects, but this should never be attempted by a beginner unless it is some simple part as it requires a great deal of study and the draftsman must practice for a long time. As this book is written with the idea of giving the beginner sufficient knowledge to work in drafting in a short time, I will not take up the subject of perspective as I consider this a subject in itself. However, I advise the beginner after mastering the general rules and securing work in drafting to get a good book on perspective and study the subject.

Sometimes shafts and other long and slender parts would extend beyond the border line of the drawing if drawn to the proper scale; therefore, it is permissible to break away as shown at shaft (1) in Fig. (3), Plate 5. It is understood that the part broken away is the same shape and size as the part shown.

CHAPTER IV

INKING THE DRAWING

Never begin inking a drawing until it has been completely finished in pencil.

Thoroughly clean the drawing before inking, carefully removing all dust, lint, etc., and see that no greasy or oily spots, or finger marks are on the surface of the paper. The best way to clean the drawing is to go over it lightly with a piece of art gum and then brush off with a soft cloth.

Shake the bottle of ink well before you start to ink the drawing. When the bottle stands undisturbed for some time the heavy part of the ink gradually settles in the bottom and the coloring of the lines will therefore vary.

After filling an instrument with ink, try it on the border of the bristol board outside of the cutting out lines, and see that your ink will flow properly and that you have the right thickness of line.

Before inking a line think what kind of a line it is, that is, whether light, medium, or heavy, and then gauge the distance between the nibs accordingly.

Start inking at the top and left-hand side of the drawing and work down and towards the right, and in this way the ink is not smeared by triangle, tee-square, or hands.

Be sure the pens are perfectly clean, and when filling the same with ink care should be used not to overload. Too much ink in the pen will make the line heavier than desired and is liable to cause blots, and too little ink, on the other hand, is liable to make a gray or streaked line.

Care should also be taken not to allow ink to get on the outside of the blades, as it is then apt to cause a blot as you draw the pen along the edge of the tee-square or triangle.

Be particular not to get lint or fuzz in the pen points as it will make a smear line.

If a number of lines radiate from a point be sure and allow sufficient time for each line to dry, otherwise it is very liable to blot.

The best results will be obtained by inking in the following sequence:

1st. Ink all large circles and arcs with the compass, beginning at the top of the sheet.

2nd. Ink all small circles and arcs with the bow-pen, beginning with the uppermost. Be sure and shade all the circles as you go along.

3rd. Draw all horizontal lines with the ruling pen and tee-square, starting at the top and left-hand side of sheet. Be sure the flat side of the pen is laid against the tee-square with the adjustable nut away from the edge. The taper of the blade is sufficient to throw the point enough so that if the pen is held vertical, or nearly so, there will be no blotting.

Hold pen at starting point for just an instant until ink begins to flow, then move with moderate speed towards the right along the edge of the tee-square. On reaching end of line pen should be immediately lifted to prevent the ink from spreading.

It is easier to draw a straight line up to a curve than to take a curve up to a line. That is the reason why all curves and arcs should be inked before the straight lines.

4th. After inking all horizontal lines the next in order are all vertical lines. They are usually drawn against the edge of the triangle, but the tee-square can be used if desired; if using triangle for guiding surface always begin at the left-hand side of the drawing and work over to the right, thus avoiding the chances of blotting.

5th. Draw with the triangles and tee-square all 30°, 60° and 45° lines, except those for cross sections. All cross sectional lines should be made after the outline of the drawing has been completed.

6th. Ink the other oblique lines which happen to be in the drawing. Use triangle for ruling surface as it is the most convenient to handle and being of transparent material is the easiest to adjust to the lines.

7th. Ink all curved surfaces which could not be made with the compass or bow-pen. For these lines use the irregular curve. Be sure to hold the ruling pen vertical, that is, don't have it slanting towards or away from the edge of the curve. If you do you are liable to blot or cause a ragged line.

8th. Make all small connections, fillets, etc., and all irregular or broken lines. Use the crow-quill or freehand pen for this work.

9th. Ink all cross section lines, making sure all adjoining parts are sectioned in opposite directions. Be sure and make these lines finer than the outline.

10th. Shade all concave and convex surfaces if doing so will add to a clearer understanding of the invention. Be sure and consider your drawing before starting any shading. You must remember if you shade some curved surfaces you must shade all curved surfaces.

11th. Shade whatever other surfaces you consider necessary.

12th. Ink in the "Inventor," etc., which is penciled in lower right-hand corner.

13th. Ink in the border lines. The line on the right side and at bottom should be at least twice as heavy as the other two sides. See Plate 5.

Carefully look over your drawing to see you have not missed anything.

CHAPTER V

SECTIONS AND SECTION LINING

Sections.—Sometimes it is necessary to show the inner construction of an object and to show it clearly. Dotted lines are often used to show these invisible parts, but this often makes the drawing appear complicated, and makes it hard to read. To eliminate this confusion of lines an object is often represented as though it had been cut by a plane, all parts between this plane and the observer being considered as removed. This is called a section or a sectional view. Figs. 2 and 3 on Plate 5 are sectional views. If possible to show just where this cutting plane passes through the object, long broken lines are used. This is shown at X-X in Fig. 1, Plate 5, which shows where the section shown in Fig. 3 is taken.

If the section is taken in the direction of the length of the object it is called a longitudinal section.

If the section is taken perpendicular to the longitudinal, that is, at right angle to the length of the object, it is called a vertical section. Fig. 2, Plate 5, is a vertical section.

When only a small portion of a view is removed to show some special part, it is called a detail section.

Section Lining.—The solid surfaces cut by the section plane are shown by what is known as section lining or cross hatching. This is shown by drawing a series of equidistant lines on the exposed area. These section lines should be drawn at 45° angle to the horizontal

SECTION OF MATERIALS.

Wood or Metal.
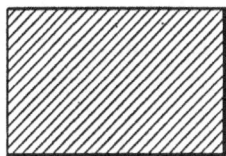

Glass or Porcelain.
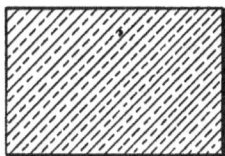

Refractory Material.
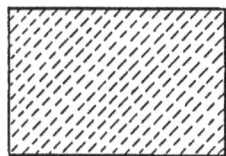

Cement.
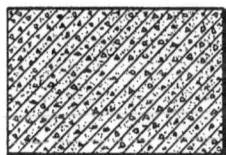

Insulation.
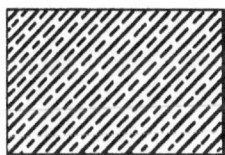

Carbon.
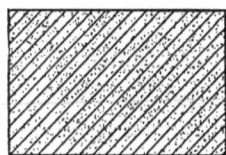

Cork.
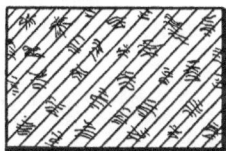

Fibrous Material.
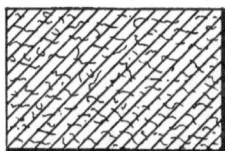

Liquid.
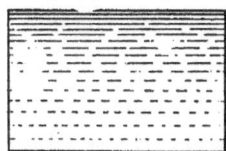

Fabrics.

Coarse
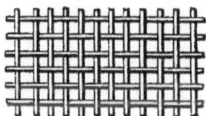

Fine
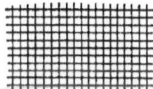

Color Chart.

Red   Blue   Green   Yellow   Black   Purple   Orange

PLATE 6

when possible, about 1/16″ to 1/20″ apart, unless it is a very small area. The beginner is very liable to space his lines too close and should be very careful to avoid this error. He should keep in mind that the drawing is to be reduced later and if too close sectioning is made, it is liable to give a solid effect.

Section lining by the eye requires considerable practice, as it is difficult to keep the lines equally spaced. After drawing six or eight lines of sectioning it is a good idea to glance back over the completed area and correct any variation in the spacing. By continually watching the lines of the completed portion you can judge whether to widen or narrow the spaces, and with practice will be able to get them uniform throughout.

Adjacent areas should be sectioned in opposite directions in order to more clearly define the limits of the parts. See parts (2) and (3), Plate 5. In case there are three or more areas adjoining the draftsman can draw the section lines at 60° and 30° angles instead of the usual 45° angle.

The section lines should be considerably lighter in weight than the lines used in the outline.

It is not necessary to cross section bolts, shafts and cylindrical solids even though they are cut by the section. This is shown in Fig. 3 where the section is taken at X-X in Fig. 1, through centre of shaft (1), but the shaft is not shown cut. For representing different materials in section a standard chart is shown on Plate 6. Also a color chart, and the proper way to make fabrics is shown on the same plate.

CHAPTER VI

OUTLINE SHADING

Outline shading is always required in patent drawings and therefore the beginner should pay particular attention to the few rules which govern the same.

The shade lines are used to make a drawing less difficult to read, and add greatly to the clearness of the views by indicating the relation of the surfaces to one another, that is, they show whether the part looked at is above or below the plane of the surface.

On Plate 7, Fig. 4, let 1-2-3-4 represent the plane of a surface. Then by the shade lines it is easy to tell that 5 is above the plane while 6 is below. Just by glancing at the figure you can readily see that 5 is a square boss while 6 is a square hole. Without the shade lines it would be necessary to draw another view of the object to show this.

The light is always assumed to come from the upper left-hand corner of the drawing and to move downward and to the right at an angle of 45', always moving in parallel lines as indicated by arrow lines in Fig. 4.

The surfaces touched by these lines, representing the direction of light rays, are always light lines, while the lines not in the light are always heavier or black lines.

All views of an object are to be considered as top views and equally exposed to light throughout their entire surface, that is, all figures whether bounded by straight or curved lines have heavier shade lines on those sides which cannot be touched by parallel lines drawn at 45° angle irrespective of their location on the sheet. Therefore, you can always use shade lines on the right hand and lower edges of all surfaces above the plane. Below the plane is just the reverse.

Shade lines should be about two or three times the thickness of the ordinary outline. There is no fixed rule for this and much depends on the drawing. In some cases if the lines were inked as heavy as this they would merge into other lines, or in case the views shown were very small it would make them appear clumsy.

In shading straight lines the extra thickness of the line is usually added to the outside of the line. This is done generally by placing the edge of the tee-square or triangle slightly below the line to be shaded, then the pen is opened a trifle wider and a line is ruled parallel to the same. The handle of the pen is then slightly inclined towards the draftsman and by drawing a line in this position, the space between the two lines will be filled. For most shading of straight lines, however, the pen can be set wider and the shade line ruled so close to the outline as to make it blend all into one thick line. Some draftsmen, when inking the shade lines, simply widen the nibs of the pen to the desired thickness and draw the one line.

To shade a circle, first ink the circle with a light line. Then with the same radius and a centre located slightly eccentric and on a 45° angle with the centre, draw another semi-circle, adding thickness to one side or the other, depending upon whether it is above or below the plane of the surface. Above the plane is shown at 8, in Fig. 4, Plate 7. Below the plane is shown at 7.

For all concentric circles the same eccentric centre is used for shading. This is shown at 7 where a represents the concentric centre and a' the eccentric centre.

SHADE LINES.

Fig.4.

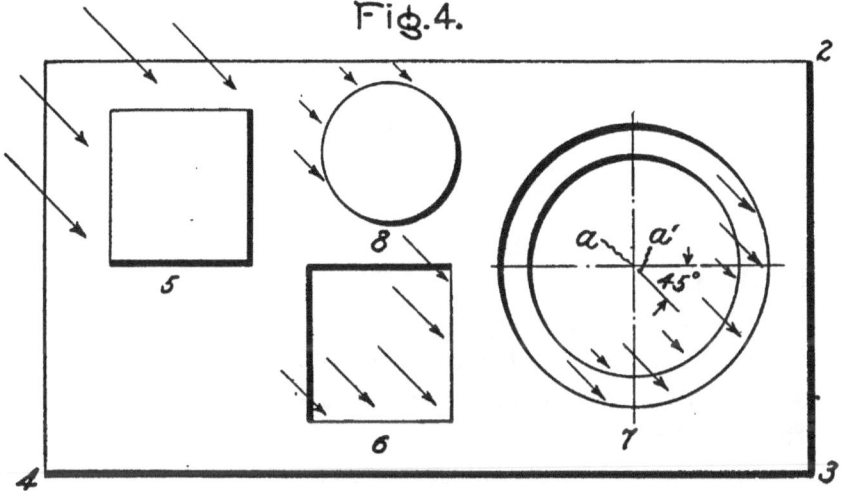

SURFACE SHADING.

Fig.5.

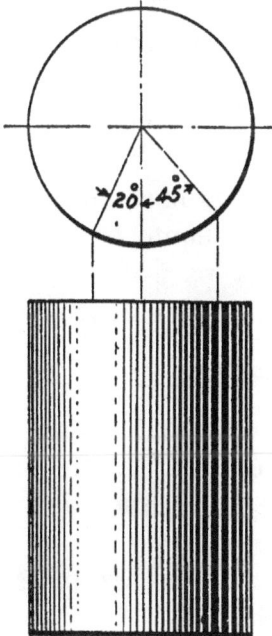

Fig.6.

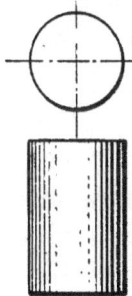

Fig.7.

Fig.8.

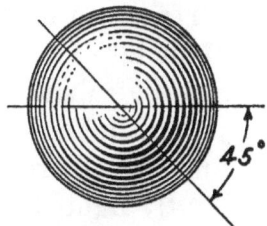

Fig.9.

Fig.10.

Fig.11.

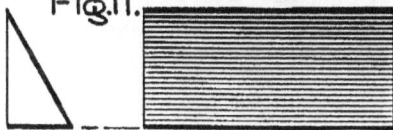

PLATE 7

CHAPTER VII

SURFACE SHADING

Shading of the surfaces of objects is not required by the Patent Office, but it is desired in a great many cases in which its use makes the reading of the drawing much easier. All patent draftsmen are called upon some time to use shading; therefore, it is a good idea for all beginners to learn how to shade, and then to practice considerably, as it requires extensive practice and extreme care in order to obtain pleasing results.

Concave and convex surfaces are almost always shaded and are shown by drawing parallel lines. On large convex surfaces the heavy and light parts of the surface are often found by drawing a semi-circle the diameter of the cylinder and then drawing a line at an angle of about 20° to the vertical centre line and projecting to the cylinder surface. The point where it touches will represent where the surface is most brilliantly lighted and therefore where there should be no shading. In the same way the darkest part of the surface can be found by projecting from a line drawn at 45° to the centre as shown clearly in Fig. 5, Plate 7.

Medium convex surfaces shown by Fig. 6 are more generally shaded by starting from the edges and gradually decreasing the thickness of the lines and at the same time increasing the space between the lines as they approach the centre.

The lines on the shaded side of the surface are of course made heavier than the other side.

Small convex surfaces as in Fig. 7 are always shaded on the heavy side only.

In shading concave surfaces, see (12), Plate 5, the operation is just reverse, that is, the darkest surface is opposite side to that shown in convex.

Large spheres are usually shaded by drawing concentric circles and arcs, the point of brilliancy, that is, the point where no shading is shown, being considered at a point about two-thirds distance from the upper edge to the centre on a line drawn at 45° with the centre and upwards to the left.

The darkest part is about one-third distance from lower edge to the centre on the same 45° line. See Fig. 8.

Most spheres, however, are usually shaded with the heaviest lines at the edges and gradually decreasing the thickness and increasing the space between the lines. As the centre is approached the arcs are also gradually shortened as can be readily seen in Fig. 9. Sometimes small spheres are shaded on the one side only as in Fig. 1, Plate 5.

Flat surfaces are shaded by spacing lines an equal distance apart. This produces what is called a flat shade. This shading is used when several surfaces in different planes lie so close that it is not easy to distinguish their location. This is shown in Fig. 10.

Inclined surfaces, as in Fig. 11, are usually shaded by parallel lines evenly spaced, the thickness of the lines being gradually reduced as the bottom of the incline is approached. The greater the angle of inclination the closer the spacing of the lines should be.

Surface shading should be used very sparingly and should not be used at all unless the draftsman is sure it will add clearness to the drawing by making it much easier to read and understand.

The Patent Office does not require surface shading nor does it encourage its use except for such cases where the drawing cannot be readily understood without its use.

CHAPTER VIII

LETTERING THE DRAWING

In filing patent applications, a description of the drawing is always required. This is usually written by an attorney, and the description will refer to the different views by figures and to the different parts by letters or numerals. The different figures should be consecutively numbered, and the different parts, when they appear in more than one view of the drawings, should always have the same reference letter.

The attorney will usually place the numerals or letters on the drawing with pencil and the draftsman will be required to put them neatly in ink and connect to the parts to which they refer with short broken or wiggley lines, as shown on Plate 5.

They should not be placed on shaded or sectioned surfaces, and when it is difficult to avoid this a blank space must be left in the shading or sectioning so the number will stand out clearly as shown at 1 in Fig. 2 of Plate 5.

No fixed style or size of lettering is required, but the Patent Office does distinctly specify that all letters and figures must be carefully formed. They also require that the letters and figures should be at least ⅛″ high. If there is sufficient room so that the letters will not appear crowded they should be made larger, as the drawing is reduced to about 6″ x 9½″ when the patent is issued, and if made too small the letters will be indistinct.

CHAPTER IX

SKETCHING

It is very important that a patent draftsman should be able to sketch at sight different objects or things he may see, such as bolts, shafts, pieces of machinery, etc. It very often happens a patent drawing is required of some piece of mechanism which is so located that it is impossible to work with drawing instruments and it is then necessary to make sketches of the parts. These free-hand drawings should be so clear that on the return of the draftsman to his office, or even at some later date, he will be able to make a patent drawing which will show clearly and concisely the mechanism.

To sketch rapidly and clearly requires practice and only with practice can anything in this line of work be accomplished.

It also requires a thorough training in observation, which in itself is one of the best educational processes.

It is very necessary to train the eyes to proportion. It is sometimes impossible to take any measurements, and it is then necessary to assume the size of a certain part and then compare the other parts to be drawn with this part. The scale is unimportant, but proportions should be secured as nearly as possible.

Perspective or isometric views are often useful as a substitute for many views, but if unfamiliar with this method, the orthographic representation will suffice for all cases.

First find that part of the mechanism which is new and sketch each detail making it complete in itself. Then make a general outline of the whole device.

Sketch centre lines, and proportion with the eye the distances of the parts from these centres.

Sectioned lined paper is good to use in this work.

Pencil should be H or HH grade. Never use a hard pencil.

Draw the lines lightly at first and correct any portion necessary, then when considered right make the lines heavy.

CHAPTER X

CARE OF INSTRUMENTS

The drawing instruments are of such a delicate construction that, unless good care is taken of them, they will become worn or broken and will not work properly. After using an instrument it should be wiped clean and laid away in the case. It should never be thrown down carelessly on the desk or into a drawer as the points are liable to be damaged.

About every two or three months the threads on all small screws should be oiled. If this is not done the threads will soon become badly worn or stripped and the instruments will be useless. Use a good grade of oil, applying with a small feather or broom splint. Be careful not to put on too much oil.

The pens become worn with continued use on the hard, calendered surface of the bristol board and therefore must be sharpened in order to secure fine, smooth and clean-cut lines. Some draftsmen send their pens to be sharpened to the manufacturers who make them, or to the dealers who sell them, but the majority prefer to sharpen their own, as very little time is required for the operation. Extreme care must be used, however, as it is a very simple matter to spoil a pen by not examining it frequently during the process to see how the points are shaped.

All sharpening should be done on a fine, close-grained oil stone. If possible, it is best to secure one which is flat enough to go between the blades of the pen when they are screwed apart, so that if during the sharpening process a small burr should form on the inside of the blades it could be easily removed. No sharpening should be done on the inside of the nibs, however, as this would spoil the pen.

The first requirement for a well-sharpened pen is to make the blades exactly the same shape and length. To do this, screw the blades of the pen together until they touch. Then, holding the pen perpendicular to the face of the oil stone with the thumb screw directly towards or away from you, draw the blades across the stone, tipping first to the right and downward, then back to the perpendicular, and then to the left and downward. Repeating this operation a few times will not only make the blades the same shape and length but will also make a well-rounded point. However, this process makes the points dull and you will be unable to make fine lines until the edges are sharpened. For this part of the operation you proceed as follows:

By means of the thumb screw open the blades wide. Rub one of the blades lightly on the oil stone, directly from and then towards you. Hold the pen at an angle of about 15 ' with the surface of the oil stone and gradually increase the angle during the process until an angle of 30° is reached. As you rub the blade to and from you, give it a slight twisting movement so as to keep the point well rounded. Of course, the same operation is required for the other blade. Frequently examine the blades and when the points appear to be fairly sharp, screw them together and draw both fine and heavy lines. If the lines are not smooth and clean cut, sharpen some more until they are. Be careful not to get the points too sharp as they are liable to cut the surface of the bristol board. If the ink does not flow easily rub very lightly on the inside of the blade, as perhaps a small burr has formed.

CHAPTER XI

Blue Printing, Etc.

The rules of the U. S. Patent Office require all drawings to be filed in Washington, and when they are once filed they cannot be returned. Therefore, it is always necessary to make some kind of a copy for future reference. Some patent attorneys simply have their draftsmen make a pencil tracing on thin paper, which they put in their files to use later when perhaps some change is required. However, most attorneys use the blue printing process, for with this means of making copies any number can be secured at very little cost. I will therefore describe this operation.

Blue printing is the means of duplicating by means of the action of light on sensitized paper. This paper, called blue-print paper, is usually furnished the draftsman in sheets the size of the patent drawing.

A frame for holding the drawing and blue-print paper is required. This frame has a glass front, and a removable back which is held in place by spring clips.

Before starting the operation of blue printing, the bristol board on which the drawing is made should be made transparent as it is too thick for the light to pass through easily. This is done by placing the drawing in a tray of gasoline and allowing the bristol board to absorb the fluid. When the paper is saturated take it from the tray and allow it to drain.

Remove the back from the blue print frame and place the drawing on the glass with the inked side touching the glass. A sheet of the sensitized paper, which must be perfectly dry, is placed upon the drawing with the yellow side (sensitized) touching the drawing. Smooth the paper out so that it will lie perfectly flat on the drawing, and fasten the back in place with the clips. While this is being done the paper should be kept from the light as much as possible. The frame is then placed where the sun's rays can shine on it directly. If convenient, artificial light such as electricity can be used.

The length of time required for printing varies according to conditions, that is, whether sky is clear or cloudy; or, if artificial light is used, varies directly according to distance from light and strength of light.

The best plan is to expose a number of small strips for different lengths of time until the desired color is secured and then record the time allowed.

The print, having been exposed the correct length of time, is taken from the frame and placed in a tray containing clear water. The print should be removed in some dark part of the room, away from any window. It should then be placed in the tray with the yellow side down, and care should be taken to have every part covered with the water. Let the print soak for about six or eight minutes, then take by the upper corners and lift out of the water. Dip it back again and repeat the operation until it appears to get no lighter. Then remove and either hang up or lay on blotting paper to dry. The operation can be repeated for as many prints as desired.

Some attorneys prefer to keep for their records a negative from which prints can be made at any time.

This negative is secured by photographing the drawing on bromide paper, and from the bromide any number of positive prints can be secured by blue printing as before described.

CHAPTER XII

USEFUL HINTS

1. Never fill the pen too full.

2. Always try the pen on the margin around the drawing to see that you have the desired thickness.

3. Never lay your pen or other instruments away without thoroughly cleaning them.

4. Do not press heavily on the ruling pen as it will cut the surface of the bristol board.

5. Do not press heavily on the pencil, so that if necessary the line can be erased and no indenture will be left in the surface of the paper.

6. Always ink your section lines after all the outline is completed.

7. Be careful not to get ink on the outside of the blades of your ruling pen. It is sure to cause a blot as you draw it along the edge of the tee-square or triangle.

8. All drawings should be made with a pen. Never try to use a brush.

9. Never use anything but black India ink.

10. Do not use both hands when locating the needle point of compasses. It looks awkward. The only exception is when the lengthening bar is used, and then it is necessary to employ both hands.

11. The ink, when it will not flow, may sometimes be started by drawing a piece of paper between the ends of the blades.

12. Sometimes just moistening the end of the finger and touching it to the points will start the ink.

13. Keep the threads on your instruments well oiled. This will save wear on the threads and make the screws easier to adjust.

14. Never ink any part of a drawing until the drawing has been entirely penciled.

15. Be particular to have the legs of the dividers exactly the same length and the points very sharp so that holes made in the paper when spacing are very small.

16. Keep the drawing board away from heat or a damp place as it is liable to warp.

17. Never allow a coating of dry ink to form on the points of the pens. Always wipe clean when through using.

18. Keep the drawing paper in a dry place and be sure it is laid perfectly flat. It is a good idea to keep it under a press or weight.

www.ingramcontent.com/pod-product-compliance
Lightning Source LLC
Chambersburg PA
CBHW022028190326
41519CB00010B/1630